What do Grampa Nelson and I do?

One sentence says it.
We do things together!

Today we pack a picnic basket.

We spot a rustic, hidden place.

We like rabbits and insects!

We like to get lots of mullets.

So what do we like best?
One sentence says it.

We just love to talk and discuss things!

Understanding the Story

Questions are to be read aloud by a teacher or parent.

1. Who is this story about? (a boy and his grandfather)
2. What do they pack? (a picnic basket)
3. What animals do they see? (rabbits, insects)
4. What do Grampa Nelson and his grandson like best? (to talk and discuss things)
5. What word on page 2 means the opposite of "alone"? *(together)*

Brown Publishing Network, Inc.
Editorial: Marie Brown, Gale Clifford, Maryann Dobeck
Art/Design: Trelawney Goodell, Virginia Pierce, Taurins Design Associates, NYC
Production: Joseph Hinckley

© Saxon Publishers, Inc., and Lorna Simmons

All rights reserved. No part of this publication may be reproduced, stored in a retrieval system, or transmitted in any form by any means, electronic, mechanical, photocopying, recording, or otherwise, without the prior written permission of the publisher. Address inquiries to Supervising Copy Editor, Saxon Publishers, Inc., 2450 John Saxon Blvd., Norman, OK 73071.

Printed in the United States of America
ISBN: 1-56577-857-X
Manufacturing Code: 01S0402
1 2 3 4 5 6 7 8 9 10 BBA 06 05 04 03 02

Phonetic Concepts Practiced

Schwa
(Nels*o*n, hidd*e*n, rabb*i*t)
vć|cv (basket)
vc|cv́ (discuss)

High Frequency Words Practiced
Decodable:
sentence

Grade 3, Fluency Reader 6
First used in Lesson 7